ISBN 978-1-5277-4105-8
PIBN 10886224

1 MONTH OF
FREE
READING

at
www.ForgottenBooks.com

By purchasing this book you are eligible for one month membership to ForgottenBooks.com, giving you unlimited access to our entire collection of over 1,000,000 titles via our web site and mobile apps.

To claim your free month visit: www.forgottenbooks.com/free886224

A STUDY OF
THE SOURCE AND CHARACTER
OF THE GASES FOUND IN HOT WATER
HEATING SYSTEMS

BY

WILLIAM CALVIN ADAMS

———

THESIS

FOR THE

DEGREE OF BACHELOR OF SCIENCE

IN

CHEMISTRY

———

COLLEGE OF LIBERAL ARTS AND SCIENCES

UNIVERSITY OF ILLINOIS

1915

UNIVERSITY OF ILLINOIS

May 25, 1915.

THIS IS TO CERTIFY THAT THE THESIS PREPARED UNDER MY SUPERVISION BY

William Calvin Adams

ENTITLED A Study of the Source and Character of the Gases

Found in Hot Water Heating Systems.

IS APPROVED BY ME AS FULFILLING THIS PART OF THE REQUIREMENTS FOR THE

DEGREE OF Bachelor of Science in Chemistry

H. J. Broderson

Instructor in Charge

APPROVED: *W. A. Noyes*

HEAD OF DEPARTMENT OF Chemistry

CONTENTS.

	Page.
Introduction.	
Historical	2
Literature on the subject	2
Object of experiment	2
Experimental- Part I. Analysis of gases	
Methods and procedure	3
Results	5
Discussion	7
Character of systems examined	7
Theories of corrosion	8
Experimental- Part II. The formation of gases	
Methods. and procedure	15
Results	18
General discussion	19
Summary	22
References	23

ILLUSTRATIONS.

———

                                                        Page.

Figure I. Apparatus used in experiments with
                iron and water   16

Figure II. Burette used to measure small
                quantities of gas   16

INTRODUCTION.

History and Object of Research.- It has been known for a great many years that gases collect in the radiators of hot water heating systems. The only apparent inconvenience that is known to result from these gases is a stoppage of the circulation of the water when the volume becomes too large. In such cases the valves near the top of the radiators are opened and the gases allowed to escape. On account of the generally accepted idea that these gases are atmospheric oxygen and nitrogen, with perhaps a few other gases, not much was thought about their presence here. It was only by accident, or perhaps for the sake of curiosity, that some one placed a lighted match to the escaping gas, and much to their surprise, found it to burn with a long blue flame. Even with this knowledge of the gas as an impetus to further investigation, nothing was done to determine its composition. It was thought that the occurance of this combustible gas wasonly an exception and, as a general rule not present.

The lack of knowledge concerning the composition or even the occurence of this gas was very evident when a search of the chemical and engineering literature was made. In discussing the theories of corrosion and their application to boilers at a meeting of the Western Society of Engineers in 1909(1), this question of the occurence of a combustible gas was brought to their attention. Dr. Burgess, formerly of the University of Wisconsin, replying to the inquery, suggested that it was probably hydrogen and this is the first authentic state-

ment as to the probable composition of the gases.

In a letter (2) addressed to the editor of Domestic Engineering, a very interesting account was given about a new Honeywell hot water heating system, in which a combustible gas had been formed in the radiator after the plant had been in operation a short time. The editor, in reply, spoke of a radiator on the third floor in his own home which failed to circulate when a fire was started in the fall. The air valve was opened, and on account of the peculiar odor coming from the escaping gas, a lighted match was applied to it. A blue flame, probably eighteen or twenty inches long, shot out from the valve with a slight roar due to the pressure of the water filling the system. The same occurence was noticeable once before when the system was one year old. The editor advanced several theories as to the probable cause of the formation of the gas. First, that hydrogen was produced from the water in some manner. Second, that the breaking up of some light mineral oil, probably present in the pipe joints, was the source. And third, that the decomposition of the vegetable matter in the water might cause it. In his opinion, the light mineral oil theory was the one most probable, and in accordance with it, gas would collect as long as oil was present in the system. He also stated that the kind of system had nothing to do with the formation of the gas. There were no authentic references given nor has anything been found in the literature to substantiate these statements. The important point in this article, however, was the presence of the combustible gases.

It was suggested that the investigation of the

composition and formation of these gases would prove to be a
very interesting subject, and might throw some light upon the
theories of corrosion. It was with these objects in view that
the investigation of the gases in hot water heating systems was
begun, and in this paper, the results obtained, together with
any bearing they may have on the theories of corrosion, will be
given.

EXPERIMENTAL

PART I - ANALYSIS OF GASES.

Method- The collection and analysis of the gases seemed to be the most logical place to begin this work, and on account
of the close connection between it and technical gas analysis, the
apparatus used will be found to be practically the same. The
glass sample tubes were of the ordinary oblong type, capable of
holding 500cc. A water jacketed burette, using water as the
confining liquid, and accurate to .1%, was used for measuring the
gases. The Hempel double absorption pipettes and the explosion
pipette were used in the determination of the different constituents of the gases.

In collecting a sample of gas, the sample tube, filled with
distilled water, was connected to the valve of the radiator by a
short rubber tube, and the valve opened. Since the expansion
tank in a hot water system is located above the highest radiator,
there is sufficient pressure to force the water out of the tube
and displace it by gas. The stopcocks were closed and the
sample brought to the laboratory for analysis. The carbon dioxide was first absorbed by passing the gas into a 50% solution
of potassium hydroxide. The heavy hydrocarbons were next re-

moved by absorption with a solution of fuming sulfuric acid, containing 20 % free sulfur trioxide.   Oxygen was absorbed by
alkaline pyrogallol solution, and carbon monoxide was removed by
ammonical cuprous chloride solution.   In determining the combustible gases, about 10 cc. of the residue was mixed with 20 cc.
of 95 % oxygen and the mixture exploded in an explosion pipette
in the usual manner.   The contraction and amount of carbon dioxide formed were measured, and assuming that methane was the only
hydrocarbon present, the amount of hydrogen in the original gas
was calculated by the equation:

$$H = 2/3 \ (\text{ contraction } ^- 2 \text{ x carbon dioxide formed.})$$

The amount of methane in the gas was equal to the amount of carbon dioxide formed in the explosion.

In a few cases, the percentage of combustible gases was
not high enough to explode; so that about 5 cc. of hydrogen was
introduced, and the correction made in applying the formula.
It was thought that the error in introducing hydrogen was probably very large; therefore another method was tried.   In this
method, about 50 cc. of the residue was mixed with 20 cc. oxygen,
and the mixture passed over palladium black in a U-shaped tube,
causing the union of oxygen and hydrogen by catalysis.   By
keeping this reaction at 100°C., the reaction between the oxygen
and hydrogen only took place, leaving the hydrocarbons unchanged.
After all the hydrogen was removed, the methane was determined by
the explosion method.   Satisfactory results could not be obtained by this method, and it was, therefore, discarded.   In cases
where only small amounts of combustible gases were present, it
was found much more satisfactory to introduce hydrogen and ex-

plode the mixture as described above.

After the analysis of the gases,it was thought that there might be some interesting results obtained by examining the water. Samples were collected from three different systems, and the water tested for free carbon dioxide and bicarbonates. In every case, the water was found to react alkaline to phenolphthalein. This shows that no free carbon dioxide was present. A .1 normal solution of sulfuric acid was used in making the titrations. First phenolphthalein was used as the indicator , and acid added until the color disappeared; then methyl orange added and the titration continued until the color change was noted. Either free alkali, normal carbonate or bicarbonate may be in the water, and it is impossible to draw any definite conclusions from these titrations. However, they show in a general way the condition of the water in the systems.

Results- In the following tables, a condensed form of the data is given.

Titration of the Water in the Systems

Location of System	System Number	Acid with Phenolphthal.	Acid with Methyl O.
911 Oregon	1C	.5Ccc.	3.05 cc.
7C8 Goodwin	100	.10	5.63
1CO7 Oregon	200	.10	5.9C

Results of Analysis of Gases in Radiators.

Location	No	Time Examined	Age System years	Analysis						Remarks
				CO_2	H.H.	O_2	CO	H_2	CH_4	
911 Oregon St.	11	Dec. 4	6	0.1	0.0	0.4	0.0	2.9	17.5	Same water in system for 2 years... Down
"	12	Jan 14	6	0.9	0.5	0.4	0.2	0.0	17.0	stairs Radiator.
619 Indiana Av	20	Feb 18	4	0.3	0.1	0.1	0.1	44.4	22.1	
916 California	30	Feb 23	3	3.9	0.1	2.3	0.2	76.2	8.9	
"	31	Mar 7	3	0.2	0.1	18.4	0.0	14.1	0.5	
708 Goodwin	101	Nov 20	3	0.2	0.1	0.4	0.3	0.0	1.2	
"	102	Dec 2	3	0.3	0.6	0.1	0.6	4.2	10.1	
1007 Oregon St.	201	Jan 12	New	0.4	0.2	0.4	1.0	74.6	3.5	
"	202	" 15	"	0.1	0.0	2.5	0.9	61.2	3.7	
"	203	" 15	"	0.1	0.3	1.7	0.0	72.0	1.5	
"	204	Feb 14	"	0.3	0.0	0.6	0.0	87.8	3.3	Collected during 24 hours.
"	205	" 14	"	0.1	0.5	19.0	0.0	5.0	3.1	
"	206	Mar 17	"	0.2	0.1	0.4	0.2	67.9	1.8	

DISCUSSION.

A brief description of the systems will give a better understanding of the results shown above. The work of collecting gases was begun in the fall, as soon as the furnaces were started. On account of the mild weather, the temperature of the water did not go very high, and so gases were not to be found in many radiators. About fifty systems, located in the Cities of Champaign and Urbana, were examined with the result that only five were found to contain gases. It must not be understood that these systems never contained gases , for in many cases the radiators had been opened and the gases allowed to escape. Some systems were closely watched to see if any gases collected, but even in these cases there was a possibility of a leak in the radiator or the valve. It was very noticeable that the radiators on the second floor often contained the only gases in the system. The method of connecting the radiators to the boiler was examined to see if an explaination could be given to account for this peculiarity. It was thought that possibly the place of formation of the gases could also be located, but nothing definite was found that would throw any light upon either of these questions.

It is very interesting to notice the effect of heat upon the production of gas. For example, in system 200, the gas was removed and about 24 hours later 900 cc. of gas was drawn off, sample 204 being some of this gas. The weather during this time was very cold and the water had to be kept almost at the boiling point. This same system was examined when the weather was warm, and only a small amount of gas was obtained.

Sample 205 was taken at another time, when the weather was warm, and the analysis of this showed that air had been admitted to the system, probably thru the contraction of the water in the expansion tank.

The length of time that the systems have been in use vary from three months to six years. In comparing the composition of the gases with this classification in view, it will be noticed that the new system, 2C0, does not contain more than 3.7% methane, while the hydrogen content runs high. In the oldest systems, Nos. 1C and 2C, the methane is high but the hydrogen is low. This might be explained by assuming that the coating of rust increases as the system gets older, and this coating acts as a protective covering to the iron, thus lowering the amount of hydrogen formed. The absorbable gases are low in all cases except samples 31 and 205, where air was probably admitted to the system according to the explanation given above. In summing up these results, the percentage of methane runs as high as 22.1, and is present in all gases, particularly in the old systems; while hydrogen runs as high as 87.8.%, but is not present in all gases. Duplicate analyses were run on all determinations given above.

The statement has been made in the first part of this paper that a few theories had been advanced as to the cause of the formation of these gases. Since the composition has already been determined and the radiators examined, it is now possible to discuss the probable cause of the formation in a more intelligent manner. All the theories that can be applied to the formation of combustible gases in a hot water heating system will be considered.

Probably the first theory of any importance regarding

the corrosion of iron is known as the acid theory. Calvert (3)
and Brown (4) were early workers who determined many facts relating to this theory. Since an acid is the chief factor in the
corrosion of iron according to this theory, and since carbon
dioxide is present in nearly all water, this acid is supposed to
be the corroding agent. This reacts with the iron forming a
ferrous carbonate or perhaps a soluble ferrous hydrated carbonate,
$Fe_2H_2(CO_3)_2$, and the hydrogen liberated combines with any dissolved oxygen, until this element is entirely used up, when the
hydrogen will be given off. However, if the solution containing
the dissolved salt is exposed to the air, it will be converted
into hydrated oxide or rust, liberating carbon dioxide, which
will attack more iron. The whole support of this theory is
based upon the presence of a free acid, but since corrosion takes
place in a faintly alkaline solution of a mineral salt of the
alkali metals, such as sodium nitrate, the presence of a free
acid is not essential to corrosion. The fact to be recognized
is that corrosion takes place in the presence of an acid, whether
free of combined, assuming that an iron salt forms before the
precipitation of rust.(20)

         The theory now generally accepted by chemists is known
as the electrolytic theory, proposed by Whitney in 1903.(5)
According to Friend, this explanation is very easily applied to
corrosion. The following quotation from this author gives a
very clear statement of the theory. (6)

    " Whitney regards the whole subject as an electrochemical
    one, and the rate of corrosion being simply a function of
    electromotive force and resistance of the circuit. Water
    is assumed to be an electrolyte, that is to say, a small

proportion of its molecules are ionized, yielding equivalent
amounts of hydrogen and hydroxyl ions."

" If a strip of iron be now introduced into such
a system, a minute quantity passes into solution with the
formation of iron ions, an equivalent amount of hydrogen
ions losing their electric charges, whereby they are converted into atoms, and are precipitated upon the surface
of the metal as a thin film of free hydrogen gas. In
solution, therefore, we have virtually ferrous hydroxide,
as is evident from the following scheme:-

$$Fe + 2H_2O = Fe(OH)_2 + H_2$$ "

Some very interesting experiments performed by Lambert[7]
and Thompson confirm this theory and show that carbon dioxide
is not an essential factor in corrosion. This theory has gain-
ed rapidly during the last few years, and all evidence seems to
point to its confirmation.

Jewett, Dustan and Goulding[8] found that iron in contact
with pure oxygen and water, in the absence of carbon dioxide,
would rust, which verifies the results obtained by Whitney and
others, but they explain the corrosion of iron by a theory of
oxidation promulgated by Traube[9] in 1885. According to this
theory, hydrogen peroxide is formed as an intermediate product
as shown in the following equation:

$$Fe + OH_2 = FeO + H_2$$
$$H_2 + O_2 = H_2O_2$$

The peroxide thus formed reacts further with the iron oxide to
form rust. Although hydrogen peroxide has been detected in the

oxidation products of many metals it has never been found in the case of iron. Dustan thinks that although it was not found, it does not disprove its momentary formation. Several investigators have brought forth evidence which opposes the peroxide theory, among these are Moody(10), Dony(11), and Divers(12).

In all the experiments relating to the corrosion theory, iron in its compact form only has been dealt with. The reaction between iron powder and water was first studied by Hall and Guibourt.(17) Ramann (18) found that 10 grams of iron powder liberated 12 cc. of hydrogen from boiling water in one hour. Because of the corrosion of the glass vessel after being heated for several days, it was thought that the glass might have some catalytic action. It was with this point in view that Friend(19) placed 10 grams of pure iron in a copper flask with boiling water. Even under these conditions it was found that a steady stream of hydrogen was given off, but the evolution was much faster when glass was present.

Many biologists have tried to explain the corrosion of iron by assuming that the lower forms of life play an active part in the reaction. There are many facts concerning the corrosion of iron that may be associated with the action of a living organism. (13) The fact that some salt solutions prohibit rusting, and others encourage it; that oxygen, water and carbon dioxide are necessary; that rusty nails cause blood poisoning; all of which support this theory. It has also been pointed out that certain organisms live in some organic ferrous salts, and that iron disintegrates very rapidly when exposed to the forces of nature, as in decaying matter. Although the biological theory

is not to be overlooked, corrosion does take place in the absence
of these organisms, as has been proven by Friend.(14)

In summing up the theories of corrosion, the evidence
seems to eliminate all but two explanations. According to the
electrolytic theory, the presence of pure oxygen and liquid water
alone are essential to the formation of rust; while the acid
theory requires an acid in addition to the other substances.
In either case, the formation of hydrogen gas as one of the
products of the reaction is the interesting fact in connection
with this work.

Directly connected with the formation of hydrogen by
the reaction of water and iron according to the corrosion theories,
is another theory which involves the reaction of water on the
compounds in the iron to form a combustible gas. One of the
chief constituents found in iron is carbon. Although the amount
of this element is very small, the proportion and the form in
which it occurs are very important in determining the character-
istics of the iron. Since free carbon in iron gives no reaction
with water or acids, it will not be discussed; but as a carbide
of iron, its properties are very interesting. One definitely
known carbide has the formula Fe_3C, and is separated from the
iron by dissolving it in very dilute sulfuric acid. The action
of water on this carbide at ordinary temperature is very slight,
and at 145° C., only .5 cc. of gas is formed by heating one gram
of carbide with 5 cc. of water. (21) At 80°, this action with
dilute hydrochloric acid produces an abundant evolution of
hydrogen. Another compound of carbon , found particularly in

hardened steel, although found to a certain extent in all kinds
of iron,is known as hardening carbon. The nature and composition
of this compound is not yet known, but it is believed by many
prominent chemists that it is carbon contained in a definite
modification of iron carbide,Fe_3C. (22) On dissolving iron in
dilute hydrochloric acid, this carbide reacts to form a combustible gas. At higher temperatures other acids will attack it
giving the same gas.

Probably there is no connection between the gases found
in radiators and the carbon in the iron, for the action of water
on these constituents is nil, and the only conditions under which
any gas will form is when an acid is present to decompose the
carbide. Cast iron from which radiators are made, contains about
4 % carbon, and there is a possibility that combustible gases will
form if water in a heating system is sufficiently acid to react
upon the carbide at the temperature of the radiator. As the
water in the systems was found to be alkaline, there is not much
probability that the carbides are acted upon.

In the smelting of iron, large amounts of gases are
dissolved in the molten iron. These gases are dissolved according to the same laws that apply to solution of gases in liquids,
and upon solidification, large amounts are held mechanically.
According to Müllers experiments with dissolved gases, many
interesting facts have been found.(15)   Molten iron dissolves
carbon monoxide, carbon dioxide, nitrogen, hydrogen and a very
small quantity of oxygen, but these gases are either decomposed
or given off , to a certain extent, and the only gases in any
amount in the cooled metal consists of hydrogen and nitrogen.

It has been found that hydrogen is retained in Bessemer steel in
larger quantities than any other gas. A very valuable experiment
was performed by Roberts — Austin,(16) in which he prepared very
pure iron from ferric chloride and when this iron was heated with
water at 70° C., hydrogen was evolved for several hours. At this
temperature, the evolution of gas ceased, but if it were placed
in a porcelain tube, in the absence of water, and the temperature
raised, hydrogen would be given off even at 1300° C. However,
if the metal after cooling is heated up again hydrogen will be
evolved, but not as rapidly as before. There can be little doubt
but that the hydrogen is held in a state of occlusion. In applying these facts to the formation of gases in radiators the quantity
of gases formed would be very small and the only combustible gas
formed would be hydrogen. This is the last of the theories or
possible sources of the formation of gases, in which iron plays
an active part.

There are, however, other ways in which combustible
gases may be formed in the radiators, and probably the most important one involves the grease used in the joints of the pipes.
The grease may be heated to such a temperature as to cause it to
break up into simpler hydrocarbons of it may react with the water.
This is a very possible source of some of the combustible gas,
but depends upon the kind of oil used.

Another possible source by which these gases could get
into the system is thru the water. The water may and usually
does contain gases dissolved in it, or it may contain organic
matter which when heated decomposes into gases which burn.

EXPERIMENTAL

PART II - FORMATION OF GASES.

Method - Now that all the possible sources, whereby
combustible gases could be produced from any part of a hot water
heating system have been discussed, it is necessary to find out
if possible from which source or sources the gas actually comes.
In order to do this the conditions in the radiator were duplicated
as nearly as possible and the theories tested, eliminating all
conditions except those bearing on the particular theory in
question.

The apparatus consisted of glass stoppered wash bottles
of about 500 cc. capacity, as shown by A Fig. I. Tube (a), which
is connected with the top of the flask is closed with a screw
clamp. (b) is a long tube coming from the bottom of the flask,
and is connected to (c), by a rubber tube. The top is sealed
with sealing wax to prevent leakage. The great disadvantage to the
these bottles was that the glass was so thick that it could not
withstand the long heating, and would crack in a short time.
The next bottle used was an ordinary Jena Florence flask of the
same capacity, but stoppered with a thoroughly cleaned rubber
stopper. B, in Fig.I, is an illustration of the flask and shows
its position when in use.

On account of the small quantities of gases to be analyzed in these experiments, a special burette, shown in Fig. II,
was constructed for this purpose. (A) is a glass water jacket
surrounding the leveling tube (L), and the glass stopcock burette
(B). Connected to (B) and (L) by the long rubber tube is the
leveling bulb (C), which contains the mercury used as the

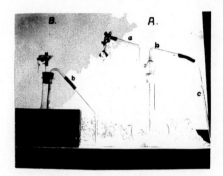

Fig. I

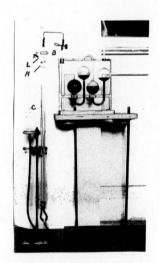

Fig. II

confining liquid.   The burette has a capacity of 30 cc. and is graduated to .05 cc.   so that by estimation .01 cc. could be measured.   The other apparatus and reagents used in these experiments were those used in ordinary gas analysis.

The interior of the radiator presents a surface which is hard to duplicate on a small scale.   Granulated cast iron of a known composition was therefore used in the glass flasks. Before placing the samples of iron in the flasks, they were thoroughly agitated with ether and alcohol to remove the adhering grease, and then dried on the water bath.

In getting the boiled and aerated distilled water, special precautions were taken to remove as, much carbon dioxide as possible.   Air was passed thru a strong solution of potassium hydroxide and then thru concentrated sulfuric acid and finally thru the distilled water to be saturated with air and purified from other dissolved gases.   This was continued for twenty-four hours, and the water either boiled for about an hour to remove the air or used in the aerated form.   In the experiments where tap water from the university mains was used, care was exercised to get a representative sample.

The flasks, after being filled, were placed in the water bath as shown in Fig. I, B, with the end of the long tube (b) immersed in water in the beaker.   The length of heating varied from one to four weeks, and the temperature of the water ranged from 60° to 90°C.   After heating the required length of time, the gases were measured and analyzed.

Results - The results obtained from these experiments are given below.

Experiments With Iron and Water

No	Kind of Water	Vol H₂O	Other Substances	Iron No	Gram Mass Iron	Time Heated Days	Vol Gas	CO₂	H.H.	O₂	CO	CH₄	H₂	Remarks
10	Boiled Distilled	500	—	6851	5	13	8 cc							Sample Lost
11	" "	500	19 KNO₃	6851	5	7	0.0							
12	" "	500		Elec	3	8	5.45	.97		.183	0.0	0.0	52.0	
13	" "	500	"	"	4	7	15.85	1.2		.81		0.0	72.5	
14	" "	500	Grease			7	0.0							
20	Aerated "	500		6850	5	25	26.25							Sample Lost
21	" "	500		68.50	5	7	41.0	.49	0.0	2.3	0.0	0.0	59.3	
22	" "	500				9	3.75	0.0		5.3	0.0	0.0	0.0	
30	Tap	400		6851	9	7	80.4	.31		.37	.62	9.5	28.5	
31	"	400		6851	9	13	18.2	.55		.22		.97	42.9	Expm 30 continued, Gas replaced by distilled H₂O
32	"	500		6850	5	13	62.0	0.0		6.5	.2	7.1	43.0	
33	"	4500				.7	81.0	4.3	0.0	1.9	.9	10.3	11.3	
34	"	4500				34	55.8	1.1	0.0	0.0	0.0	16.2	0.0	
35	"	500				9	5.88	0.0	0.0	.51	0.0	26.2	0.0	
40	Boiled Tap	500		6851	6	13	40.3	1.0	0.0	.6	.7	2.5	73.0	

19

Analysis of Iron

Lab.No	Total C	Graph.C	Comb.C	Si	Mn	S
6850	4.03 %	3.93 %	.10%	2.56%	.363%	.032%
6851	4.25	3.80	.45	2.03	.229	.017

GENERAL DISCUSSION

Duplicates were not run on either the water and iron experiments or the analysis of the gases. The volumes of the gases, even when produced under apparently the same conditions, seemed to vary widely. This can be explained by the fact that the temperature was not kept constant and the gases may not have formed or may have dissolved again before the analysis was made. The discussion of the results will be taken up in the order in which the theories of the formation of the gases were mentioned. Experiments 10, 20, 21, and 22 were run for the purpose of showing the relation between the production of gas in the presence or absence of an acid, such as carbonic acid. No. 10 shows that there is some action between cast iron and distilled water, and Nos. 20 and 21 show that the presence of air greatly aids the production of gas while 22 gives a check on the amount of air dissolved in the water. According to the acid theory, even a small amount of carbon dioxide will cause the solution of a large amount of iron if oxygen is present. Even in boiled distilled water, there is probably enough carbon dioxide present to start the reaction and if oxygen is introduced the reaction will be greatly increased. Although the electrolytic theory

does not require carbon dioxide to be present to produce a solution
of the iron, it does require oxygen to accelerate it.   Since some
of the hydrogen produced according to these theories is supposed
to unite with some of the dissolved oxygen, gases might$_\wedge^{not}$ be given
off.   This, however, is not the case here, for in each experiment
with iron and water large amounts of hydrogen were given off.
It appears that oxygen does increase the corrosion, but that it
does not combine with all the hydrogen.   A series of experiments
has just been completed by Cobb and Dougill (24), in which the
dissolved oxygen has been found to play an important part in the
corrosion of iron.

Since the iron was in the form of fine shavings, some
other factors may enter into the reaction which are not present in
the reaction between solid iron bars and water.   These results,
however, agree with those obtained by Ramann (18) and Friend (19).

It was thought probable that the dissolved salts in the
system might have some action upon the iron in such a way as to
produce a gas.   Experiment 11 was carried out for this purpose,
but since Heyn and Bauer (23) have shown that different salts have
different effects upon iron and that the concentration is the
essential factor, this experiment does not definitely prove
anything.

To test out the carbide theory, Nos. 10, 12, and 13 were
prepared.   No. 10 contained boiled distilled water and cast iron,
while Nos. 12 and 13 contained boiled distilled water and filings
of electrolytic iron which had been melted and cooled in a vacuum.
Gases were given off in each case showing that there seemed to be
no difference between pure iron and cast iron.   This experiment

also shows that occluded gases do not enter into the reaction to a very large extent, if at all.

The grease theory was next tested. A sample of grease was obtained from one of the large plumbing shops of Champaign, and placed in a flask with boiled distilled water. No gas was produced, showing that the gas could not come from this source. These results are in agreement with those of other investigators who have studied the reaction between grease, water and iron.(25)

Up to this time it will be noticed that no methane has been found in any of the gases given off by the action of iron in water. It is also interesting to note that no tap water has been used. Experiments 30, 31, 32, 33, 34, 35, and 40 were run for the purpose of determining the effect, if any, that tap water has in the formation of combustible gases. With the exception of 34 and 35, both hydrogen and methane were formed to some extent. It is certain that the amounts of dissolved gases will vary from time to time, but the presence of such a large amount of hydrogen in No. 33 has not yet been accounted for.

The university wells are about 175 feet deep, so there is, therefore, a possibility that methane from deposits of natural gas, would get into the water. The other possible source, already mentioned, is that the methane may come from the decomposition of organic matter. It would seem that the amount of organic matter would have to be rather large to give such a large amount of methane. It is very probable that the organic matter has something to do with the formation of this gas, for the volume of the gas gradually increases as the heating continues.

It would seem that if methane is formed by either of

of these methods that after heating a short time, the gas would all be driven off. In system 1C, the percentage of methane is very high, and yet the water has not been changed for two years. Likewise we would expect a large amount of methane in a system when fresh water is heated for the first time. System 2C0, however, shows the contrary to be true. The only explanation which seems plausible is that the amount of the other gases given off in proportion to the methane is very large at first, but as the quantity of these decreases either by continual heating or by a removal of the cause of formation, the percentage of methane will increase. If this explanation is accepted, the production of methane has to come almost entirely from the organic matter.

Although the exact source of methane in the radiator is not known, it seems evident that its presence is due to the tap water and that the hydrogen is produced by the action between the water and the interior surface of the heating system.

SUMMARY

I. Gases were found to collect in almost all hot water heating systems.

II. The quantity and composition of the gases vary within wide limits, depending upon the temperature of the water and the length of time the system has been in use.

III. The combustible gases were found to be methane and hydrogen.

IV. Hydrogen is probably produced by the action between the water and the iron.

V. Methane is probably produced from the tap water, either by the decomposition of the organic matter or from the evolution of the freviously dissolved gas.

23

REFERENCES.

1. Boiler Corrosion as an Electrochemical Action.
   Jour. Western Soc. Engineers. Vol. 14, p 375.

2. Trouble with Gas in Radiators.
   Domestic Engineering. Vol. 66, p 108.

3. The Acid Theory of Corrosion. Grace Calvert.
   Chemical News. 1871. Vol. 23, p 98.

4. The Acid Theory of Corrosion. Crum Brown.
   Jour. Iron and Steel Inst. 1888. II p 129.

5. The Oxidation of Iron. W. R. Whitney.
   Jour. Am. Chem. Soc. 1903. Vol. 25 p 394.

6. The Corrosion of Iron and Steel. J. N. Friend.
   Page 36.

7. Wet Oxidation of Metals. Thompson and Lambert.
   Jour. Chem. Soc. Part I Vol. 97, p 2426.
   Part II Vol. 101 p 2056.

8. Effects of Salts and Other Substances on the Process of
   Rusting. Dunstan, Jowett and Goulding.
   Trans. Chem. Soc. Vol. 87, 1548.

9. Traube. Ber Vol. 18, 1981.

10. The Rusting of Iron. G. T. Moody.
    Jour. Chem. Soc. Vol. 89, 720-30 1906.

11. Uber die vermeintlicke Radioaktivitat des Wassers
    stoffsuperoxide. Dony and A. Dony Chem. Zentb. P II, 203.

12. Divers. Proc. Chem. Soc. Vol. 21, 251.

13. The Busting of Iron. H. Richardson.
    Nature. Vol. 74, 586. 1906.

24

14. The Corrosion of Iron. J. N. Friend.

   Jour. Iron and Steel Inst. Part II, 258.

15. The Constitution of Iron Alloys and Slags.

   H. F. von Juptner.   Page 284.

16. Fifth Report to the Alloys Committee of the Inst of

Mech. Eng.   1899.   Roberts-Austen.

17. Gmelin- Kraut.   III 300.   Hall and Guibourt.

18. Zersetzbarkeit des Wassers durch metallisches Eisen.

   Ramann.   Berichte. Vol. 14, 1433.

19. The Busting of Iron.   J. N. Friend.

   Jour. Iron and Steel Inst. 1908. Part II, 5.

20. The Corrosion of Iron and Steel.   Friend.

                              Page 36.

21. Siderology. H. F. von Jüptner.   Page 165.

22. Siderology. H. F. von Juptner.   Page 173.

23. Uber den Angriff des Eisens durch Wasser und wasserige

Losungen.   Heyn and Bauer.   Mitteilungen aus den koniglichen

Materialprufungsamt, Berlin.   1910, V. 28,62; V. 26, 2.

24. Corrosion by Dissolved Oxygen. J. W. Cobb and G. Dougill.

   Jour. Soc. Chem. Ind.   Vol. 33, 403.

25. Corrosion of Iron and Steel. J. N. Friend.

                        Page 172.

Printed by BoD™in Norderstedt, Germany

9 781527 741058